Bradley Tice

Statistical Physics and the Fundamentals of Minimum Description Length and Minimum Message Length

GRIN Verlag

Bibliografische Information der Deutschen Nationalbibliothek:

Die Deutsche Bibliothek verzeichnet diese Publikation in der Deutschen Nationalbibliografie; detaillierte bibliografische Daten sind im Internet über http://dnb.d-nb.de/ abrufbar.

Imprint:

Druck und Bindung: Books on Demand GmbH, Norderstedt Germany
ISBN: 978-3-656-35143-6

This book at GRIN:

http://www.grin.com/en/e-book/205420/statistical-physics-and-the-fundamentals-of-minimum-description-length

Statistical Physics and the Fundamentals of Minimum Description Length and Minimum Message Length

Bradley S. Tice

Abstract

This monograph is the first account of the use of a 'summing engine' an algorithm for counting and unifying common, or liked natured, characters in a binary sequential string and combining them into a sub-group of common character types into a compressed collective.

The ability to compress a random binary sequential string is a novel feature of this 'Summing Engine' algorithm and is examined in light of both Minimum Description Length and its progenitor Minimum Message length that use data compression as a parameter for defining 'good' data for evaluating a model for measurement.

A new paradigm for both Minimum Message Length and Minimum Description Length results from this study.

Preface

The application of a counting, or addition, algorithm to compress common, or liked natured, characters in a binary sequential string that reduces, into a compressed state, a random binary sequential string and then utilized in both the Minimum Message Length and the Minimum Description Length models as test systems for the traditional random state and the non-traditional random state as developed by the author in 1998. A new paradigm results changing the fundamental notions in both the Minimum Message Length (1968) and the modern version of the Minimum Description Length (1978) model.

Dr. Bradley S. Tice November 2012

The Table of Contents

Introduction

The monograph will address the 'compressibility' of a traditional random binary segmental string verses a 'summing engine' binary sequential string. The model used for both compression systems are the Minimum Message Length, MML, and the Minimum Description Length, MDL, with the result being a paradigm shift of both the Minimum Message Length and to the Minimum Description Length systems at the fundamental level.

Minimum Message Length, MML, was first developed by C. S. Wallace and D. M. Boulton (1968). The bases of Minimal Message Length is very similar to Minimum Description Length except the Minimum Message Length is a fully subject Bayesian model (Wikipedia, 2011: 1 &4).

Minimum Description Length, MDL, was first developed by J. Rissanen in 1978 (Rissanen, 1978). The bases of the Minimum Description Length is that regularity in a specific set of data can be used to compress the data into a sequence shorter than the original length of the originating sequence.

Minimum Message Length

Minimum Message Length [MML] was the early progenitor of Minimal Description Length [MML] and was first published by Wallace and Boulton in 1968. The primary difference between MML and MDL, the Minimum Description Length, is that the Minimum Message Length is a fully subjective Bayesian model in that it is of a 'a prior' distribution (Wikipedia, 2012:5-6).

In the MML model all the parameters are encoded in the first part of a two-part code so all the parameters are learned (Wikipedia, 2012: 5-6).

Minimum Description Length

Minimum Description Length [MDL] was developed in 1978 by Jorma Rissanen as a method by which the "best hypothesis for a given set of data is the one that leads to the best compression of the data" (Wikipedia, 2012. 1). Minimum Description Length has had to bypass two fundamentals of Kolmogorov Complexity in that Kolmogorov Complexity is incomputable and uses 'what' computer language is used (Wikipedia, 2012: 2). Minimum Description Length restricts the codes allowable so that it does become computable to find the shortest code length available and that the code being 'reasonably' efficient (Wikipedia, 2012: 2).

The Minimum Description Length Principle is as follows:
The best theory minimizes the bit sum of the length of the description of the theory and the length in bits of the data when encode by the help of the theory (Li and Vitany, 1997: 351). Minimum Description Length try's to balance the regularity and randomness in the data using the best model; the one that uses regularity in the data to compress (Li and Vitanyi, 1997: 351)

The Grammar of Form

Martin-Lof (1966) has developed a form of algorithmic complexity based on patterns within sequential strings of binary data of both a random and non-random manner (Martin-Lof, `1966). A more regular pattern of a binary sequential string would look like this:

[1010101010]

This type of regularity has regularity, a balance, of form of both [0's] and [1.s] that marks it as a non-random sequential string (Martin-Lof, 1966). A random sequential string would look like the following:

[0100011000]

These patterns represent the patterns of the grammar of statistical randomness and are the tradition measure of 'the qualities and quantities' of the notion of statistical randomness in a sequential string.

A Compression Engine

A 'compression engine' is, in essence, an algorithm for common types of character's ,either all [0's] or [1's], in a binary sequential string to compress into like-natured characters and then be de-compressed when desired (Tice, 2009). Much of the 'ornamental', visual markers, found in my first large scale address of a 'summing engine' is removed from this publication to save time and extraneous notation (Tice, 2009).

The 'summing engine' is a systematic process of the sequential addition of common, or liked natured characters. In this case the binary [1's] and [0's] of minimally weighted semantic values that have the [1's] being the opposite value of the [0's].

A New Paradigm

If a 'summing engine' is used on a binary sequential string the following three results will occur:

The pattern of the binary sequential string will be of a regular pattern or non-random distribution.

The pattern of the binary sequential string will be of a non-regular pattern or of a random distribution.

The pattern of the binary sequential string will be of both a regular and a random distribution.

Along with the three types of binary sequential string types, the fundamental properties of 'a priori' and 'a posteriori' mark the pre-algorithm and the post-algorithm models.

A traditionally compressed binary sequential string would be non-random as follows:

[1010101010]

A traditional binary sequential string would not compress as follows:

[0111100011]

If a 'summing engine' is used as the algorithm both non-random and random binary sequential strings would both compress as follows:

Non-random compression:

[1010101010]

Compressed:

[10] five times.

Random compression:

[1000110000] with [1x1] [0x3] [1x2] [0x4] or [1010]

If the 'summing engine' is used along with the traditional notions of Kolmogorov Complexity a great deal of change is noted for the 'summing engines' ability to compress a random binary sequential string as to the traditional Kolmogorov Complexity random model.

Minimal Description Length avoids assumptions about data generating procedures were as Minimum Message Length 'represents a Bayesian framework' (Wikipedia, 2012: 4). It has been noted in the literature that some researchers feel that Minimum Description Length is equivalent to Bayesian inference but is renounced by Rissanen as being 'data that does not reflect the 'true' nature of whether the data as collected or that data to 'reality' (Wikipedia, 2012: 4).

Because two forms of randomness exist traditional, unable to compress randomness, and the compressible type of a 'summing engine', able to compress a random sequential string, the very fundamentals of both the Minimum Message Length and the more developed model of the Minimum Description Length are divergent at the point of what makes up a 'random binary sequential string and compression'. Modern Minimum Description Length theory is based on the principle of compression, as is traditionally known; Kolmogorov Complexity, and has no development using a random binary sequential string using a 'summing engine' (Wikipedia, 2012: 1 and Li and Vitanyi, 1997: 351).

Conclusion

Using both the Minimal Message Length Model and the Minimal Description Length Model as 'tests' of randomness found in binary sequential strings against both the traditional Kolmogorov Complexity notion of randomness and the author's 'Summing Engine' form of randomness in binary sequential strings result in two very different sets of measures.

The resulting changes to both the Minimal Message Length and the Minimal Description Length models is at the fundamental level of statistical physics and adds a new chapter to the study of algorithmic complexity.

Summary

This monograph is the first account of a major fundamental change to both Minimum Message Length and Minimum Description Length. The application of a 'summing engine' to make a random binary sequential string to compress has foundational effects to the traditional notions about compression as a fundamental level of the parameters for the measure of such compression found in both the Minimum Message Length and the Minimum Description Length models.

Notes

Rissanen notes his, with Barron and Yu, that Wallace and Boulton's seminal paper (1968) to the field of Minimal Description Length as being the "crudest". Wallace and Boulton's paper of 1968 seems more 'rudimentary' than 'crude' and is, in some respects, a different set of measures than Rissanen's work(Barron, Rissanen and Yu, 1978: 1 and Wallace and Boulton, 1968).

References

Barron, A, J. Rissanen and B. Yu (1998) "The Minimum Description Length Principle in Coding and modeling". IEEE Transactions on Information Theory, Vol.44, No. 6, October 1998. Pp. 1-17.

Li, M and P. Vitany (1997) An Introduction to Kolomogov Complexity and its Application. Second Edition. New York, Springer.

Li, M. and P. Vitanyi (2000) "Minimum Description Length Induction, Bayesiansim, and Kolmogorov Complexity". IEEE Transactions on Information Theory, Vol. 46, No. 2. March 2000.

Rissanen, J. (1978) "Modeling by shortest data description". Automatica, Vol. 14, pp. 465-4711978.

Tice, B.S. (2009) Aspects of Kolomogorov: The Physics of Information. Denmark: River Publishers.

Vitanyi P. and M. Li (2008) "Minimum Description Length Induction, Bayyesnism and Kolmogorov Complexity". arXiv;cs/9901014v1 [cs.LG] 27 Jan 1999. Pp. 1-35.

Wallace, C.S. and D.M. Boulton (1968) "An Information Measure for Classification". Comp. J. Vol. 11, No. 2 pp. 185-195, 1968.

Wikipedia (2012) "Minimum Description Length". Wikipedia Encyclopedia, pp 1-6. Website: http://en.wikipedia.org/wiki/Minimum description-length

About the Author

Dr. Tice is the founder and CEO of Advanced Human Design located in The Central Valley of Northern California, U.S.A. Dr. Tice is, or has been, a member of I.E.E.E., S.I. A. M., A.M.S., I.M.S. A.C.M. and is a Fellow of The Royal Statistical Society and a Fellow of The British Computer Society.